Did you enjoy this issue of BioCoder?

Sign up and we'll deliver future issues and news about the community for FREE.

http://oreilly.com/go/biocoder-news

BioCoder #7

SPRING 2015

Beijing · Cambridge · Farnham · Köln · Sebastopol · Tokyo   O'REILLY®

BioCoder #7

Copyright © 2015 O'Reilly Media, Inc. All rights reserved.

Printed in the United States of America.

Published by O'Reilly Media, Inc., 1005 Gravenstein Highway North, Sebastopol, CA 95472.

O'Reilly books may be purchased for educational, business, or sales promotional use. Online editions are also available for most titles (*http://safaribooksonline.com*). For more information, contact our corporate/institutional sales department: 800-998-9938 or *corporate@oreilly.com*.

Editors: Mike Loukides and Nina DiPrimio	**Proofreader:** Eileen Cohen
Production Editor: Nicole Shelby	**Interior Designer:** David Futato
Copyeditor: Amanda Kersey	**Cover Designer:** Randy Comer
	Illustrator: Rebecca Demarest

April 2015:   First Edition

Revision History for the First Edition

2015-04-15:   First Release

978-1-491-92503-4

[LSI]

Contents

Revolutionizing the Conversation on Biotechnology

Keira Havens and Nikolai Braun

New technologies inevitably meet resistance. Canning, refrigeration, and pasteurization all took some time to get off the ground due to consumer mistrust. It goes beyond food—there was also resistance to both anesthesia and organ transplantation (*http://bit.ly/synbio-ethics*). These technologies all share one thing: they change the common understanding of our limitations.

In the 1800s, everyone knew that you couldn't eat vegetables stored at room temperature that were months or years old, until canning came along and you could. Up until the twentieth century, surgery was a last resort, until anesthesia made it possible to bear. Up until the twenty-first century, everyone knew that agricultural breeding was a function of skill, a keen eye, and good luck. You couldn't design a plant to do what you wanted, until you could, with modern bio-engineering tools.

The first plant genomic transformation took place in 1983. Due to the expense and complexity of the technology at that time, further genetically modified organism (GMO) development was primarily a tool of large corporations. These corporations developed crops to benefit their customers (farmers) with increased yields in commodity crops. Farmers got to spray less, till less, and reap higher yields with these enhanced crops. Farm productivity increased, consumers had more food at lower prices, and everybody won.

Except, consumers have had no idea they are benefiting from this technology.

Here's part of the problem: farmers—the people buying GMO seeds, the people who see the benefits firsthand—make up just 2% of the United States population. The proportion of people involved in farming has steadily dropped year after year as techniques, transportation, storage, and every other kind of agricultural technology that goes into growing food and delivering it to the end user improved.

Plows made from steel, refrigerated trucking, the Haber-Bosch process, and countless other innovations have steadily made the net productivity from a single acre better and made farming many acres possible with fewer personnel.

The US is now at a point where 94% of Americans (*http://bit.ly/farm-connect*) have no connection to farming or farmers, and GMOs are perceived as corporate arrogance writ large. GMOs entered the food system without consumer engagement. When consumers became aware of the technology and asked questions, they were met with the corporate equivalent of "Don't worry about it; you wouldn't understand." This approach did not inspire confidence, and today our discussions about GMOs often devolve into trite repetition of bumper-sticker opinions.

How Do We Change the Conversation About Biotechnology? We Make It Beautiful!

Computer science got its first taste of popularity through gaming, translating a string of zeroes and ones into a virtual ping-pong game. Personal computing took off when Apple took the computer from the programmers and used great design and easy user experience to bring mothers and art students on board. Technology must be made personal and beautiful before it becomes ubiquitous.

Beautiful biotechnology gives people a new palette of experiences to associate with GMO technology: appreciation, wonder, and delight. Some 80 million US households garden, making it the most popular hobby. A color-changing flower, like the one being developed by our company, Revolution Bioengineering, is a consumer biotechnology accessible to everyone who gardens, allowing gardeners to speak about GMOs from their own experience.

Others have dabbled in this game—Suntory debuted a "blue" rose *Applause* several years ago. It was a major technological achievement, but it remains illusory—a quick look at the flower shows that it is lavender at best. Glowing Plant got crowds excited about a future in which street lights are replaced by plants, but that ambitious future is still distant.

Plant biotechnology has been the subject of a lot of broken promises, and so we have set our sights on a goal that is both amazing and achievable. The flower that we are crowd funding (*http://bit.ly/color-flower*) this spring changes color on demand from white to purple over the course of a day when you spritz it with a dilute ethanol solution. We have a prototype that works now (and has a video (*http://bit.ly/flowervid*)) and have partnered with a technical team with over 25 years of petunia pigment research experience to bring it to the world.

In all other regards, this is just like any other petunia: it needs water, it makes beautiful blooms, and it dies off in the fall just like every other petunia.

Color-changing flowers will bring home the potential and possibility of this technology in a way that hasn't been possible before.

We have more-ambitious goals of course: a flower that changes color throughout the day from pink to blue and back again, flowers that are scented with vanilla, and flowers that have polka dots. But these flowers are, once again, far in the future. We have a responsibility, right now, to achieve a result that lives up to expectations and demonstrates that we can use biotechnology to make beautiful things. By embracing the beautiful, we can change the way we talk about biotechnology. We can remove biotechnology from its current context of corporate control and its unassailable savior complex. We can expand beyond the limitations imposed by misinformation and fear to discuss how this technology can be used most effectively to develop a healthier, more sustainable society.

All of these actions are necessary for us to build the trust and respect that will support the next century of biological innovations. Without that mutual understanding, we risk trapping ourselves in this fearful conversation permanently, in a frustrating regulatory environment that neither accurately reflects risk nor provides assurance for the public.

Discussions of technologies that have failed to live up to their promise routinely include GMOs. It's time to change that.

Keira Havens and Nikolai Braun are plant molecular biologists dedicated to beautiful biotechnology. They might not always see eye-to-eye, but they've learned how to share their point of view and reach a common vision. They founded Revolution Bioengineering (http://www.revolutionbio.co/) in 2013 with the goal of championing wonder and fascination in the world of science. Join them in bringing color-changing flowers to the world with their Indiegogo campaign (http://bit.ly/color-flower). They're excited to be a part of the SOLID 2015 lineup, sharing their vision of beautiful biotechnology with the O'Reilly network.

Bioinformatics for Aspiring Synthetic Biologists

Edgar Andrés Ochoa Cruz, Sayane Shome, Pablo Cárdenas, Maaruthy Yelleswarapu, Jitendra Kumar Gupta, Eugenio Maria Battaglia, Alioune Ngom, Pedro L. Fernandes, and Gerd Moe-Behrens

Abstract

For a synthetic biologist or biohacker to be able to hack, design, create, and engineer biological systems, the ability to work with biological data is essential. Basic bioinformatics skills will be required in order to read, interpret, write, and generate files containing DNA, RNA, protein, and other biological information. In this article, we will show the path you need to follow to implement a biological function using online data. As a case study, we are using Imperial College's 2014 iGEM project, which focused on the optimization of bacterial cellulose production for use in water filtration.

Introduction

There are three requirements for the design of a biological system:

- The specification of the desired system in terms of its functions, inputs, and outputs

- The use of bioinformatics skills to select and combine DNA parts that follow these specifications

- The actual genetic modification of the organism in the wet lab

The design requires more than knowing how to pick bioparts from a catalog; it implies knowing how to create them and how to combine them to achieve the desired system.

Proper bioinformatics skills allow you to extract information within biological data and use it to model the desired system. Therefore, they're of top importance for any good biohacker.

There are large repositories of bioinformatics tools for synthetic biologists. Our main goal here is to describe the pipeline (see Figure 2-1) that allows a biohacker to use some of these tools in order to complete a synthetic biology project. We will focus on a practical example: the Imperial College's 2014 iGEM project (*http://bit.ly/aqualose*). A video presentation of this project can be found on YouTube (*http://bit.ly/aqua-vid*).

The Imperial College team focused on the biosynthesis and optimization of bacterial cellulose production using *Escherichia coli*. This biomaterial is used in industry for several purposes. Compared with plant cellulose, bacterial cellulose has advantageous properties, such as its high purity and strength, and its special porosity, which is what interested Imperial's team in the material as a potential water filter. The group was interested in the porosity characteristics that could make it useful as water-filter material to address a worldwide issue, water contamination. With that goal, they tried to reduce the cost of the bacterial cellulose production, which is a main limitation for commercialization of traditional cellulose filters.

In this article, we provide a general guide to performing fundamental *in silico* tasks for the development of one of the bioparts needed for this synthetic biology application. A more detailed guide with command-line details can be found on the Leukippos Institute home page (*http://www.leukippos.org*).

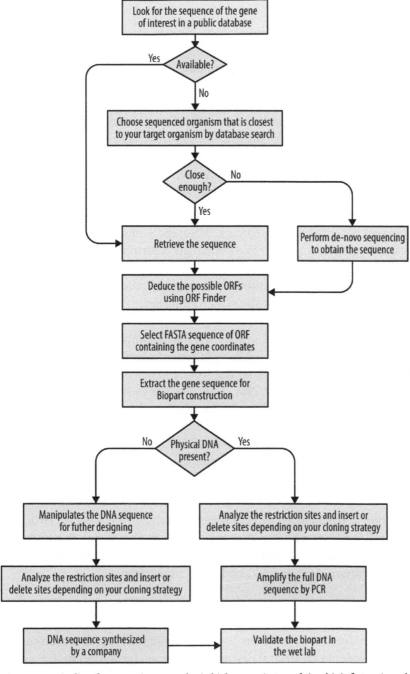

Figure 2-1. Pipeline for executing a synthetic biology project applying bioinformatic tools

From Data to Life

For this purpose, bioinformatics tools can be divided in two main groups. The first group comprises analytical tools, which help you to understand the properties of biological data. Among these tools, you can find:

- DNA and RNA folding: the visualization of predicted secondary structures or hybridization patterns formed by DNA or RNA sequences is crucial to avoid the design being ruined by the presence of unforeseen structural aberrations. These could lead to a variety of disastrous results, from unsuccessful polymerase chain reactions (PCRs) to misfolded proteins. The University of Albany's RNA Institute website (*http://bit.ly/rna-fold*) is a good place to start.

- Protein modeling: protein structure plays an important role in determining the biological functions of this biopart. Theoretical models are predicted for amino acid sequences for which experimentally determined structures are not known yet. Various techniques are employed for protein modeling, and they differ in complexity, accuracy, and the primary data required for the structure prediction. We recommend that you start with these:

    - Homology modeling: SWISS-MODEL is an automated protein structure homology modeling server that accepts requests either through the Expasy server (*http://swissmodel.expasy.org/*) or can be accessed via the eepView program (*http://spdbv.vital-it.ch/*). To perform protein modeling, the target amino acid sequence and 3D structure of a template protein are required. To determine the template protein structure, perform a BLAST (tool for searching online databases, based on nucleotide or amino acid homology) search with the target sequence against the database of previously determined protein structures (.pdb). Use the structure with maximum identity (homology) score as the template for the modeling.

    - Validation of the modeled protein structure: validate the modeled structure via the SAVES server (*http://bit.ly/ucla-saves*). The server validates the model based on six parameters and suggests if the model satisfies the essential requirements to be a good theoretical model.

The second group of tools contains the software that helps you to manipulate sequence data in order to create and design new bioparts (genes, promoters, ribosome binding sites, etc.). One could say that these are the tools that actually help you to engineer life. In this guide, we will focus on these tools to show you how to produce a biopart in your hackerspace.

Most of genetic information you need is already available online and is free to access. Among the databases that you can use, the most popular ones are:

- GENBANK, a database of gene sequences hosted by the U.S. National Center for Biotechnology Information (NCBI) (*http://www.ncbi.nlm.nih.gov/*).

- UniPROT (*http://www.ebi.ac.uk/uniprot*), a database that provides a clustered sets of sequences from the UniProtKB and selected UniProt Archive records to obtain complete coverage while hiding redundant sequences.

- ENSEMBL (*http://www.ensembl.org*) or UCSC (*http://genome.ucsc.edu/*), the main genome browsers with integrated knowledge bases. Though you can find species-specific genomic databases with a quick Google search, depending on the organism of interest, you'll find them much more reliably here.

For synthetic biology, we suggest the Registry of Biological Parts (*http://parts.igem.org/Catalog*) and the Joint BioEnergy Institute's registry (JBEIR) (*https://public-registry.jbei.org/*).

In some cases, the information is not available online, so you will need to have access to the physical DNA or organism from which you can extract the genomic DNA. Besides genomic DNA extraction, in certain contexts (e.g., gene cloning, gene probes, or creation of cDNA libraries) you might want to synthesize complementary DNA (cDNA). The cDNA is double-stranded DNA synthesized from mature messenger RNA (mRNA) by using reverse transcriptase.

Nevertheless, you will eventually need to find the sequenced organism closest to your target organism using a database search and use its sequence to design primers to amplify your biopart from the extracted DNA using PCR. The following tools can be used to design your primers, using a homologous region containing your biopart, shared by sequenced and unsequenced organisms, as a virtual template: Primer3 (*http://primer3.ut.ee/*) and Primer–BLAST (*http://bit.ly/primer-blast*).

The biopart you want to amplify will probably not be identical in both organisms; therefore, the design of degenerate primers (defined as: "a population of specific primers that cover all the possible combinations of nucleotide sequences coding for a given protein sequence" by Iserte et al., 2013 (*http://bit.ly/deg-primer*)) is recommended.

In our practical example, Imperial's team wanted to amplify parts from the *Gluconacetobacter* genome and introduce them into the *E. coli. Gluconaceto-*

bacter is characterized by its high production of bacterial cellulose but is more difficult to manipulate and scale in a biofactory than *E. coli* because of the lack of specific genetic tools and information. Therefore, the group decided to isolate and sequence a new strain, which they named *G. xylinus igem*, as well as strain ATCC 53852 in order to obtain new information. The group sequenced both bacteria to obtain the genetic information that allowed them to access the data. They also created several tools to genetically manipulate *Gluconacetobacter*. Sequencing is the third possibility for retrieving biological data, which is useful in case you don't find the data online for your organism or a closely related one. By the way, sequencing the gene of interest is always good practice to make sure you are working with what you think you are.

One Biopart at a Time

The iGEM team effort improved the knowledge and tools for genetically manipulating *Gluconacetobacter*. Nevertheless, *E. coli* is one of the most studied microorganisms in science and is easily scalable for industrial, low-cost production. Therefore, the team decided to introduce the bioparts from *Gluconacetobacter*, needed for the bacterial cellulose production, into *E. coli*.

We will focus on one of the four genes that belong to the operon (group of genes controlled by the same promoter sequence) that is responsible for bacterial cellulose production. A promoter is a piece of DNA sequence that defines where transcription of a gene by RNA polymerase begins. The methodology explained here can be applied to any biopart you are interested in. We chose the acsD gene, which is vital for control of cellulose crystallinity and high production level (see the review in *Macromolecular Bioscience* (*http://bit.ly/cellulose-paper*)).

The first step is finding the gene's sequence in the public database:

1. Type X54676 (accession number of the complete operon used by the iGEM team) in GENBANK's search bar. You will find several results belonging to protein sequences, scientific literature about this operon, etc. You need to choose the "Nucleotide" option.

2. Within the GENBANK entry, look for the acsD gene (see Figure 2-2).

```
gene              9194..9664
                  /gene="acsD"
CDS               9194..9664
                  /gene="acsD"
                  /codon_start=1
                  /transl_table=11
                  /protein_id="CAA38490.1"
                  /db_xref="GI:455537"
                  /db_xref="GOA:P37719"
                  /db_xref="InterPro:IPR022798"
                  /db_xref="PDB:3A8E"
                  /db_xref="PDB:3AJ1"
                  /db_xref="PDB:3AJ2"
                  /db_xref="UniProtKB/Swiss-Prot:P37719"
                  /translation="MTIFEKKPDFTLFLQTLSWEIDDQVGIEVRNELLREVGRGMGTR
                  IMPPPCQTVDKLQIELNALLALIGWGTVTLELLSEDQSLRIVHENLPQVGSAGEPSGT
                  WLAPVLEGLYGRWVTSQAGAFGDYVVTRDVDAEDLNAVPRQTIIMYMRVRSSAT"
```

Figure 2-2. GENBANK format showing the acsD gene

Figure 2-2 shows part of the amino acid sequence codified by the acsD gene and the nucleotide coordinates of this gene in the operon DNA (9194-9664). The easiest way to extract the acsD sequence from the complete operon sequence is to use the ORF-Finder tool (*http://bit.ly/orf-finder*). It will also help you to be sure about recovering the complete ORF (open reading frame) from the start to stop codon. Complete the following task:

1. Go to the top of the X54676 GENBANK page and select the FASTA display format.

2. Paste it on the ORF-Finder tool. Hit the OrfFind button.

3. The analysis will show you six possible frames (three forward frames and three reverse frames). Select the ORF that is in the acsD coordinates (9194-9664) by clicking the appropriate square for the frame of interest on the right-hand side: "+2 □ 9194..9664 471" (see Figure 2-3). It will also change color also on the diagram of the left-hand side.

4. Accept the ORF. Once again it will change color, indicating that it is the accepted one.

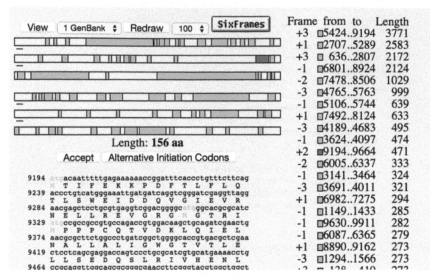

Figure 2-3. ORF-Finder format showing the acsD gene open reading frame

5. Select "2 Fasta nucleotide" instead of "1 GenBank" and press the View button. You will get the exact sequence of the acsD gene in FASTA format.

NOTE
This exercise could also have been done with the acsC, another of the operon genes. Incidentally, it would have been a bad choice because there is a mistake in the GEN-BANK sequence at 5286 position (it has a G instead of the correct A). Finding errors in the databases is not uncommon; this is why we recommend the use of the ORF-Finder tool, which will help you to be sure about your sequence and helps to detect this kind of problem in the online sequences.

NOTE
Do not forget that the ORF-Finder does not predict intron sequences. An intron is the nucleotide sequence within a gene that will not code for protein. In eukaryotic organisms, it is removed before translation of messenger RNA by the splicing process. If you have intron sequences, you can use a related gene/protein sequence and one of the tools (*http://bit.ly/softberry*).

Now that you have obtained the exact sequence of your biopart, you will need to have a company synthesize your DNA, or you can amplify out of the genome with primers and clone into a bacterial expression plasmid. For synthesized, you can ask the company to send you the DNA in a plasmid for expression, which means that the plasmid must have the other genetic elements needed for expressing the gene, such as a promoter or RBS (the ribosome binding site is the RNA sequence that must be found in mRNA for ribosomes to bind precisely and initiate translation).

Furthermore, you can modify the DNA sequence to fine-tune it for your purpose. Imperial's team optimized codons for expression in *E. coli* and used a strong RBS sequence found in the registry of parts database (BBa_B0034). They also tuned this RBS part using the Salis-Lab RBS calculator (*https://salis.psu.edu/software/*).

There are several tools to help you to modify and explore your sequence (examine the restriction enzymes profiles, insert mutations, or change the codon usage). These are graphical cloning and design tools:

- Gene Designer (*http://bit.ly/gene-designer*)

- ApE (*http://bit.ly/ape-editor*)

- Genome Compiler (*http://www.genomecompiler.com/*)

- SnapGene (*http://www.snapgene.com*)

- Benchling (*https://benchling.com*)

When you receive your biopart, you will probably want to continue with further cloning. We recommend you to plan in advance which strategy you would like to use (Gibson assembly, restrictions enzymes and ligase, etc.). We also recommend that you find which restriction enzymes don't cut your sequence. This will allow you to add these sites using primers in a PCR reaction before cloning or to ask to the synthesis company to put these sequences in your DNA. You can use NEBcutter2 (*http://tools.neb.com/NEBcutter2/*) and Webcutter 2.0 (*http://bit.ly/webcutter2*) to analyze the restriction sites.

Both have easy-to-follow instructions. It is likely that one of them is more advantageous than the other for your specific problem. You may also be happy with using the restriction site analysis that is included in one or more of the cloning tool software packages that are listed above.

Enhancing Your Capacity for Engineering Life

As the work of a synthetic biologist develops, it becomes less common to try to modify one gene at a time. As the field is growing fast, the community naturally aims at tuning dozens of genes at a time to get a desired function, while some would even want to attempt a complete genome change at once. To carry out those aims, you will need to be able to automate complex tasks into scripts that can be invoked by simple commands with little effort using command-line interfaces (CLIs). This allows for efficient analysis of large datasets while ensuring a low error rate. By "large datasets," we mean, for example, the ones resulting from the 1000 Genomes Project, which has so far produced 200 terabytes; or the ones produced by the Cancer Genome Project. For many bioinformatics applications, there is no choice but to use the command line, and it is unlikely that their authors will ever develop alternatives. Therefore, it is a good idea to be ready for that and acquire some skills to use CLI.

The use of CLI allows you to automate many tasks at your local or remote computer, such as downloading genes from the GENBANK, which already has this option implemented in the Entrez Direct (*http://bit.ly/ncbi-edirect*) tool application program interface (API). Once the tool is in a machine that you can interact with via a shell, you can download a copy of the bacterial cellulose production operon for local usage by typing a single command:

```
esearch -db nucleotide -query "X54676" | efetch -format fasta > mySeq
```

This operation will fetch a FASTA file with the nucleotide sequence in one go. esearch will get the GENBANK entry, a pipe (|) will cause efetch to pick the output and format it as FASTA, and the result will be "thrown" by redirection (>) into a file named *mySeq* in your current directory. Many variants are, of course, possible. If you want the GENBANK file format, use (-format gb) instead of (-format fasta). The GENBANK file format contains further information about the selected operon; for example, it contains the coordinates of each gene that composed it.

This type of operation can be scaled up to retrieve complete genomes, for example. Chaining such operations into complex scripts that perform a variety of analyses, prepare formatted outputs, submit jobs to remote servers, and a huge variety of operations will allow you to increase your capacity for engineering life beyond the limits imposed by simple interactive bioinformatic tools.

Conclusion

The wealth of data available online today is an invaluable resource for biohackers to tinker with. Therefore, having a minimal working knowledge base of bioinformatics is a powerful asset for designing and hacking biological systems. Whether it is RNA or protein folding, sequence analysis, primer design, or restriction enzyme analysis, there's bound to be a bioinformatics tool for the job, as we've seen with our walkthrough of Imperial College's bacterial cellulose project. We hope this introduction helped you through your first steps in bioinformatics or resolved any doubts you might have had. We also hope to have triggered your curiosity and stimulated you to acquire more elaborate bioinformatics skills to use in ambitious synthetic biology projects. Now, go hack some DNA!

Join Our Leukippos Community

If you like synthetic biology and bioinformatics, want to learn more, get some help, and be part of some great biohacking projects, join Leukippos, a synthetic biology lab in the cloud (*http://www.leukippos.org*). We have an awesome group on Facebook (*http://bit.ly/leukip-fb*). You are very welcome to join.

Donations

Our Leukippos community needs your financial support. We are all working for free and pay everything out of our own pockets, such as server costs and page registrations. If you like our work and wish that we produce more quality content, such as this paper, you can donate some bitcoins to this address: 1KnikzSG7fnRfG76DxjLZyrbvw8fS9nisw.

Correspondence can be directed to Dr. Gerd Moe-Behrens: leukipposinstitute@googlemail.com.

Edgar Andrés Ochoa Cruz (aka Don) has a PhD in biotechnology from São Paulo University, Brazil. He was the instructor of two Brazilian iGEM teams and founded Syntechbio, the first biohacker space in South America. He is managing and developing several projects in synthetic biology and developing platforms for providing home and industry users access to biotechnology. He is the cofounder of Arcturus BioCloud (http://www.arcturus.io/) in San Francisco, California, taking synthetic biology to your home safely.

Sayane Shome completed her undergraduate studies in bioinformatics from Vellore Institute of Technology in Vellore, India. She works as an external student researcher and virtual classroom trainer for bioinformatics modules at King Abdulaziz University in Rabigh, Saudi Arabia. She served as the president of RSG-India, a student body affiliated to the International Society of Computational Biology (ISCB) Student Council, from 2012 to 2014.

Pablo Cárdenas is an undergrad student at Universidad de Los Andes in Bogotá, Colombia. He has been a volunteer at the Leukippos Institute since 2012 and is interested in biology, biotech, and the ethics involved in these areas (as well as all of science) in the context of improving public health and social awareness. He is an enthusiastic supporter of open science and glad to participate in projects like Leukippos.

Maaruthy Yelleswarapu is a master's student at ETH Zurich and has been collaborating at the Leukippos Institute since 2013.

Jitendra Kumar Gupta works as a programmer and research assistant at Shodhaka Life Sciences Private Limited, incubated within the Institute of Bioinformatics and Applied Biotechnology (http://bit.ly/startbioinfo). He has a master's degree in bioinformatics from Mangalore University. For a short period of time, he worked in the Manipal Institute of Technology as a research scholar. He joined the Open Source Drug Discovery Project (http://www.osdd.net/) (an initiative by CSIR) at the Indian Institute of Science.

Eugenio Maria Battaglia is an undergraduate student in molecular biotechnology at the University of Turin with a specialization in integrative neuroscience. Currently he's developing the concept of a bio-commons license in the European framework named Synenergene (http://www.synenergene.eu/).

Alioune Ngom received his BSc degree in mathematics and computer science from the Universite du Quebec a Trois-Rivieres in 1990 and his MSc and PhD degrees in computer science from the University of Ottawa in 1994. He is a professor at the University of Windsor, Ontario, Canada. Prior to joining the University of Windsor, he was an assistant professor at the Department of Mathematics and Computer Science at Lakehead University, Thunder Bay, Ontario, Canada, from 1998 to 2000. During his short stay at Lakehead University in 1999, he cofounded Genesis Genomics Inc. (now, Mitomics Inc. (http://www.mitomicsinc.com/)), a biotechnology company specializing in the analysis of the mitochondrial genome and the identification and design of mtDNA biomarkers for the early detection of cancer. He is member of the IEEE-BBTC and IAPR-Bioinf and coleads the Pattern Recognition in Bioinformatics group at the University of Windsor.

Pedro L. Fernandes is a bioinformatics training coordinator at Instituto Gulbenkian de Ciência, in Oeiras, Portugal. He also organizes brainstorming events on challenging themes such as "Distance and eLlearning Technologies," "Systems Biology and P4 Medicine," and "Pathway Analysis in Proteomics." He is an advisor to Figshare, an ambassador to iAnn, and plays team-leading roles in organizations like EMBnet and GOBLET.

Gerd Moe-Behrens has a BSc and MSc from the University of Oslo and a PhD from the Faculty of Medicine, University of Oslo, Norway. He founded the Leukippos Institute for Synthetic Biology, a research institute solely in the cloud. Moreover, he is founder and CEO of CytoComp (http://www.cytocomp.com), a young startup focusing on biological computing.

DIY Scientific Publishing

Joshua Nicholson

Publishing for the Hacker/Tinkerer and the R01-Funded Scientist

To publish is to make something known or to disseminate it to the public. With the advent of the Internet, publishing has become very easy to do. Nowadays, anyone can set up a website or blog and publish their work and ideas immediately and freely. Indeed, literally millions of people and organizations do just that every day and every hour. The ability to publish so easily, coupled with the ability to disseminate publications through various social networks like Facebook, Twitter, and Reddit, means that you can write something on your couch and a few hours later have it be seen by millions of people. This may not be easy, but it is certainly possible. Because of this, the publishing industry has dramatically changed over recent years. Indeed, we are in a time when there are more independent publishers (bloggers) than there are publishing organizations. What can publishers offer writers that they can't really do themselves? Not much, and because of this, publishing, in short, has for the most part become do-it-yourself, except in science. Why?

In science, publishing is anything but simple. Words like "preprint," "impact factor," and "peer review" are all influential terms that affect the dissemination and readership in one way or another. For example, you can publish something as a "preprint," but that's not considered a "publication" by scientists because it has not been peer-reviewed (even if it has!). You can publish something in a low-impact journal, but that publication is seen as less than one in a high-impact journal, despite the actual impact it makes. The Web has not yet leveled the playing field in science publishing because of measures and metrics that scientists and publishers have introduced and maintained in order to uphold some semblance of value that traditional publishers offer. This is beginning to change, however, as various article-level metrics are introduced and recognized as being influential on publishing, ultimately, putting focus on the content and not the publisher—a great move for science! Indeed, the value that traditional publishers offer is becoming recognized for what it is: a brand, just like Nike. But brands have little if any effect on whether or not an experiment is right or wrong, impactful or not. Indeed, in physics and math, there are many "preprints" in the arXiv repository that have changed entire fields. This attitude that where you publish doesn't matter is spreading to biology and other fields as well. This is not to say that traditional science publishers don't offer anything to scientists—they do.

Science publishers offer standard ways that work can be cited and preserved, namely digital object identifiers (DOIs, a unique alphanumeric string assigned to a website) and archival services like CLOCKSS (*https://www.clockss.org*). These tools are now becoming available to individuals through organizations like The Winnower (*http://bit.ly/winnow-doi*) and figshare (*http://figshare.com/*) that provide DOIs and archival services to individuals. Users can now self-publish their work easily and quickly with the same tools that *Science* or *Nature* offer for thousands of dollars. Arguably, they can do it more efficiently and transparently as well.

How Does It Work?

At The Winnower, users can upload a variety of documents, such as Word, LaTeX, and even content directly from blogs (*http://bit.ly/win-pub*) for submission. These documents are immediately and automatically converted into HTML and displayed online at The Winnower, where readers can write open reviews of the content. Based upon reviews received, users have the option of updating their initial submissions and when satisfied can then assign a DOI to their work and archive it permanently. Because the content is automatically typeset and reviews are crowd-sourced, publishing can done much more cheaply and quickly than many other publishers can offer. This model has not only allowed different levels of scientists, from the undergraduate student to emeritus professor, to contribute but has also encouraged different types of content to be published. Indeed, scientists on The Winnower and figshare have published work directly from their electronic lab notebooks (*http://bit.ly/win-note*), individual datasets (*http://bit.ly/beer-graph*), blogs (*http://bit.ly/win-blog*), crowd-sourced letters written on Google documents (*http://bit.ly/win-letter*), presentations (*http://bit.ly/fox-dna*), and even posters (*http://bit.ly/101-innovations*). It's not just the types of people and content that have changed under this model; it's also the interactions among scientists. As one author on The Winnower notes (*http://bit.ly/ice-thoma*):

It leads to helpful dialogue. All of the reviews on The Winnower are as open and accessible as the articles…Because of the open nature of this review process, reviewers need to maintain a high standard (their name is attached to it!) and trend towards more constructive (rather than negative) feedback. (However, evidence for this assumption is lacking).

We hope this shift will continue to gain momentum so that science can be judged not by where it is published but by the content of the publication. That goal can be accomplished by breaking down the major publishers in science, but it can also be accomplished by building up the individual publishers (i.e., you!). We at The Winnower have chose to do the latter.

Joshua Nicholson is the founder of The Winnower, a DIY scholarly publishing platform launched in May 2014. He is also currently a PhD candidate at Virginia Tech studying the role of the karyotype in cancer initiation and progression in the lab of Dr. Daniela Cimini. He received his bachelor's degree in molecular, cell, and developmental biology at UC Santa Cruz in 2008 and is the grandson of the late UC Berkeley professor Richard Strohman, who taught him that "science is for the surprises, not the prizes." He hopes The Winnower will help uncover those surprises.

The Gap Between the Stars

Ryan Bethencourt

Elon Musk announced that he and SpaceX want to colonize Mars with millions of people within the next 20 years. While Dawn (*http://dawn.jpl.nasa.gov/*), NASA's probe, is orbiting the dwarf planet Ceres, NASA plans future probes to dive deep into Europa looking for signs of life, as well as a space submarine to navigate Titan's methane seas. India is taking great leaps toward its first manned space mission, in a race with China and Europe. Things look bright for space exploration, yet something is missing.

Robotic exploration is accelerating, but biology is being left behind. Space is a hostile place for biology; no gravity, no atmosphere, the risk of decompression, and that's just the start! The Earth's atmosphere and magnetic field shield us from 99.9% of the cosmic radiation in the open black void of space.

As we look to the stars, life may have found fascinating and surprising places to live. Perhaps there are new forms of life, as of yet unimagined, that exist in subterranean soil on Mars or deep in the oceans of Europa, waiting for our first robotic emissaries.

And What Happens When We Find Life?

To traverse the biologically hostile expanse of space, either we'll have to shield ourselves from radiation through extraordinary means (think massive masses), or we will have to figure out how to understand and aggressively cure every conceivable form of cancer because in space, it'll be a daily risk. Applied, interdisciplinary biology is woefully underfunded, and with stagnant NASA, National Science Foundation (NSF), and National Institutes of Health (NIH) funding in the US, it's unclear where the money for black skies research will come from.

Humanity stands on the verge of glimpsing at least some of the variety of life that likely exists throughout our own solar system, but before we do, we need to understand so much more about our own bodies. Why is it that mammals can't reproduce without embryonic defects in microgravity environments? Will our future space-faring descendants be infertile? How will magnetic fields and varying light levels of other planets like Mars affect us? How will we eat and grow food in zero or lower gravity environments? Can we fight the muscle- and bone-wasting effects of microgravity with biotech? Can we use biology to evolve new biomaterials on Mars? How about Venus or Europa? How do we recycle our waste or our atmosphere when most of our controlled environment experiments, like the Biosphere 2, have failed?

If we one day want to see humanity as a space-faring civilization, we need to imagine, fund, and build early-stage biotechnologies that can be used both in space and here on Earth. Stanford University, Brown University, and NASA's iGEM work on Hell Cells (*http://bit.ly/hellcell*). They installed and tested five parts from various organisms for base, desiccation, cold, and radiation resistance with the goal of building an organism that could grow and survive on Mars (and perhaps one day provide food and breathable atmosphere for us).

Terraforming alien environments is only just the start. Bell Biosystems (*http://www.bellbiosystems.com/*), Blue Turtle (*http://blueturtlebio.com/*), and other biotech companies have begun looking at reprogramming human bodies (through new synthetic organelles) and microbiomes (through genetically modified gut flora) to produce medicines and increased resistance to terran diseases that will have applications as we start to encounter harsher biological environments. We are shifting to programmable "smart" cells that can respond to new environments that our biology has yet to encounter. The only way to bridge the biological gap is to build it!

May we live in interesting times!

Ryan Bethencourt is the program director and a Venture Partner at Indie Bio, an accelerator for early stage biotechs. Ryan was previously the life sciences head at the XPRIZE Foundation, cofounder and CEO of Berkeley Biolabs and cofounder of Sudo Room and Counter Culture Labs.

Ryan has worked with most major biopharmaceutical companies in the U.S., E.U., and Japan over the last decade to develop novel drugs from first IND submission to FDA approval.

Ryan's work has been featured in Wired, TechCrunch, Forbes, Fast Company, and other publications. You can connect to him via twitter @RyanBethencourt.

Biotech Startup Advice: How to Reduce Costs, Save Time, and Increase Happiness

Tom Ruginis

Running an Efficient Lab

You've been funded, and now it's time to build a company. Yay! You might be starting with $25,000, $100,000, or $1.2 million. Whatever it is, you'll be spending 12–30% on lab supplies and equipment. Do you want to manage this complicated process? A startup called HappiLabs created the Virtual Lab Manager to help.

We specialize in the management of purchasing and finances for scientific research labs. Many biotech founders with a science background come from academia, where there is an absence of finance education. How are scientists expected to manage large amounts of money with no training? Accounting, negotiating prices, and purchasing is handled by the university or seasoned lab manager, which makes life easy.

Welcome to startup land...where you're on your own. Your team is now responsible for purchasing, negotiating, accounting, and building a relationship with suppliers for plates, PCR tubes, reagents, antibodies, media, pipette tips, and the list goes on.

Your startup will create accounts with at least 15 suppliers, and someone will be responsible for interacting with sales and customer service reps. Twenty-two minutes here and 8 minutes there, company time will be lost as you obtain pricing information, place orders, and follow up on shipping estimates (ETAs). Expect every order to consume at least 35 minutes of your time, broken up over two to three days as you handle seven tasks associated with every order.

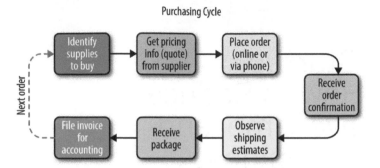

Figure 5-1. Seven steps associated with every purchase

Also, if you don't have time to shop around (step 2 of the seven steps in Figure 5-1), which consumes more time, you might overpay by 30–80%.

No one wants to waste funding from an angel, accelerator, or grant, but if this is your first time starting a company, mistakes will happen. Some mistakes are due to poor planning; others occur because of disorganized suppliers, and some mistakes happen normally in the chaotic startup environment.

Even the most intelligent and organized founder will not escape error, something that makes you a better entrepreneur, right?

Here are four tips for improving efficiency in your startup, something they don't teach you in academia.

1. STAY UP TO DATE ON ACCOUNTING

Set aside at least 2 hours every week to organize receipts and analyze invoices for erroneous charges. Set up a payroll system that calculates appropriate taxes. Use $96 of your funding to setup a QuickBooks account. Taxes will be quicker at year's end. If you don't spend 2 hours per week now, you'll spend 10 hours per week in December and January when you don't have 10 hours.

Added benefit: investors like to see organized accounting.

2. DO NOT BELIEVE IN THE "DISCOUNT"

Suppliers use the word "discount" as a marketing tactic. Be careful and make sure to analyze the price from which the discount is coming, known as list price (see Figure 5-2). 30% might sound good, but if it's 30% off $950, I'd rather take 0% discount off $565.

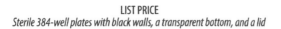

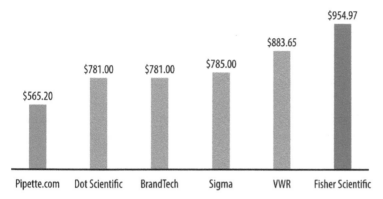

Figure 5-2. The graph shows the list price from six suppliers for the exact same item: sterile 384-well plates with black walls, a transparent bottom, and a lid. Manufactured by BrandTech (cat#781687) (http://www.brandtech.com/prodpage.asp?prodid=pureGrade)

3. PLAN FOR SHIPPING ESTIMATES TO GO WRONG

Two of the biggest headaches in your startup will be back orders and inaccurate shipping ETAs.

> *Thank you for your purchase. Your order is estimated to arrive on 3/28.*

3/29 arrives. No package. 3/30? Nothing. Now you spend time calling customer service, which may or may not have an answer. Even Amazon is unreliable sometimes. Meanwhile, scientists are sitting around, being paid, and your runway shortens with no new data.

The best advice we can give you is to be mindful and do your best to plan for shipping delays. Orders take longer than you want them to.

4. KNOW YOUR COSTS PER EXPERIMENT

Don't just buy pipette tips and taq. Buy and assign them to an experiment or protocol. Know that your assay costs $23.30 per run and takes 18 hours of a scientist's time at $x/hour, and convert that into a cost per experiment per month. Estimating the runway will be more accurate, and future investors will be delighted to know exactly where their money is going.

These four tips will save you many hours and hundreds to thousands of dollars, and increase the likelihood of future funding.

Your happiness depends on making it to the next round of funding and minimizing stress along the way. HappiLabs wants to help. We're a small company in Chicago that provides a service called the Virtual Lab Manager (*http://www.happi labs.org/*). We are specialists in purchasing and management of lab supplies and are for hire hourly to help your startup succeed. Good luck in startupland!

Tom Ruginis is the CEO and Founder of HappiLabs. He's a former molecular biologist turned entrepreneur living in Chicago. You can reach him at tom@happilabs.org or @letUbeU.

The Cost of Science: The Demand for Price Transparency

Sean Seaver

If your lab's supplies suddenly cost 30% less each month, what could you do with the money you saved?

Maybe as chemists you could buy time-saving, presynthesized chemicals instead of making them on your own. Maybe you could even hire another research assistant to help you get better results faster.

With budgets already tight, saving 30% might seem like a long shot. But when we analyzed more than 10,000 lab supply purchases, that's the amazing average. What's the biggest single factor that would allow researchers to save that amount? Transparent pricing.

We've come to expect price transparency when we're shopping for goods in our personal lives. One quick product search on Google or Amazon yields ordered lists of competing stores and prices, with shipping policies and other fees clearly marked. But when we search for lab supplies online, it's nearly impossible to find the best prices easily.

Consider this: the University of Toledo in Ohio regularly spends $8 million to $10 million on research supplies and chemicals. Saving 30% means $2.5 million to $3 million each year. For perspective, the University of Toledo increased its tuition by 3.5% in 2012 for a projected revenue increase of $7.6 million. (*http://bit.ly/ UT-tuition*)

These savings matter. So why aren't we researchers getting access to competitive, transparent pricing for our chemical and lab supplies in what is estimated to be a $39-billion-dollar-industry (*http://bit.ly/sec-vwr*)?

How Lab Supply Pricing Gets Hidden

Short answer: big scientific distributors don't want you to know a fair price, because your ignorance puts them at an advantage. They can charge whatever they think you'll pay, just as if you were buying a used car. And unfortunately, the systems that most labs use to buy supplies don't discourage hidden pricing and fees.

Here's how pricing gets hidden.

1. LAB-TO-COMPANY NEGOTIATIONS

It's standard procedure for universities and institutions to negotiate volume-based deals with large chemical distributors. Their purchasing agent typically sends out a "request for proposals" and gets back what look like great deals (see Figure 6-1).

SCHEDULE B

**MISC. Scientific Chemicals, Equipment & Supplies
Sample Pricing Proposal Form**

Discounts Offered- *The following represents a sample exhibit of how to complete the Schedule B Pricing Form.* The Schedule B Pricing Form is found in the attached Excel file.

Brand	Discount
Hewlett Packard – Chromatography Supplies	17%
HF Scientific	10%
Leica – Microscopes	27%
Panasonic	25%
Sanyo - Centrifuges	10%
Sanyo – Ovens	20%

Figure 6-1. Price proposal sheet

The problem is that the proposal prices are based on percentage discount, not absolute value. That incentivizes manufactures to increase their list price, which is why Fisher, for example, lists exorbitantly high prices on its website although no researcher in her right mind would pay them.

Purchasing agents usually don't have a scientific background, so they don't know that the prices on big distributors' websites are a joke.

These negotiated deals lead to some pretty crazy price differences from lab to lab. Just look at this: the same exact product (acetonitrile, a commonly used solvent in chemistry), same company, same day, but as offered to three different Midwestern academic research labs:

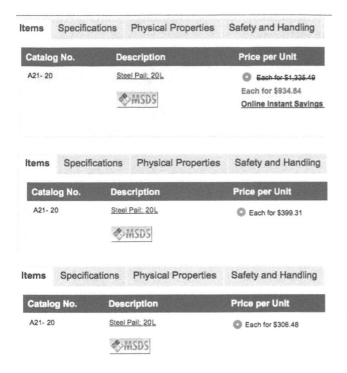

Yikes, right? We even see labs at the same institution paying different prices on the exact same products.

Here's another example. Say you need to buy 100 grams of 98% pure etomidate.

You could pay anywhere from $1,400 to $29,000 to, no joke, $1.6 million, depending on the company you buy it from. These types of differences slow if not stop scientific progress.

You can debate brands, quality, and pricing, but in this example, the purity is the same. Thankfully, the American Chemical Society (ACS) has taken measures to help standardize products by creating an ACS standard that helps researchers compare products that meet their specifications from different suppliers. The result helps make basic research chemicals commodities and like deciding which gas station to use should largely be determined by price.

2. HIDING PRICES BEHIND A QUOTE BUTTON

Notice the green button in Figure 6-2? Even with university contracts, scientific distributors still often require requests for quotes. This adds another layer of hidden pricing and creates more inefficiency.

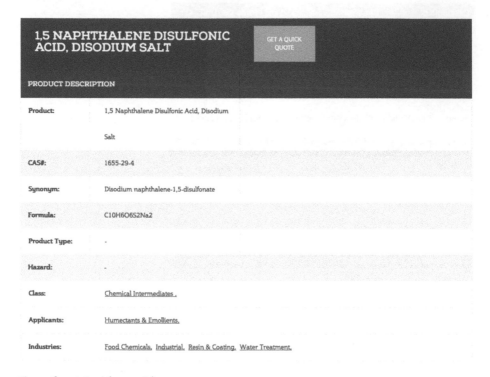

Figure 6-2. A "quick quote" button

3. PUSHING OUT SMALLER DISTRIBUTORS

Since the big distributors often negotiate a requirement to bid on any scientific products from an outside vendor, they will often quote just under what was provided by a smaller company. This results in smaller manufacturers not publishing their prices on their websites.

But calling every small manufacturer to ask about pricing is not feasible for most scientists. We're busy. We need the information at a glance.

It's too bad, because these smaller manufacturers can offer price advantages if you buy from them directly; sometimes, they're selling their product to the big distributors, which rebottle it, change the label, and sell it for 10 times more.

4.DISCOURAGING OUTSIDE PURCHASES

When making a purchase, scientists often look for items in the university purchasing system, because going outside that system (typically composed of 10 or so companies) usually requires extensive paperwork.

Scientists have to decide if a better price outside the university system is worth the time it takes to create a purchase order.

Larger companies have used this to their advantage and increased their prices. The result is some items costing 5 to 10 times what they would from ordering around this purchasing system.

5. HIDING BIG FEES

Imagine buying a $20 item from Amazon and then two weeks later getting a $65 bill for shipping. I realize this sounds crazy, but it is common practice in orders for scientific supplies and chemicals.

Figures 6-3, 6-4, and 6-5 are from various suppliers' checkouts pages.

Figure 6-3. "A nominal shipping charge may be applied"

VWR Terms and Conditions of Product Sale

1. **Acceptance** - ALL SALES ARE SUBJECT TO AND EXPRESSLY CONDITIONED UPON THE TERMS AND CONDITIONS CONTAINED HEREIN, AND UPON CUSTOMER'S ASSENT THERETO. THE TERMS AND CONDITIONS CONTAINED HEREIN WILL BE CONTROLLING, AND ANY ADDITIONAL AND/OR INCONSISTENT TERMS AND CONDITIONS SET FORTH IN ANY ACKNOWLEDGMENT, PURCHASE ORDER, OR ACCEPTANCE DOCUMENTS REQUESTED FROM AND/OR PROVIDED BY CUSTOMER ARE EXPRESSLY REJECTED. NO VARIATION OF THESE TERMS AND CONDITIONS WILL BE BINDING UPON VWR UNLESS AGREED TO IN WRITING AND SIGNED BY AN OFFICER OR OTHER AUTHORIZED REPRESENTATIVE OF VWR.

2. **Specifications** - Product specifications are subject to change without prior notice.

3. **Delivery** - Delivery of all orders will be FCA (INCOTERMS 2010) and title of all Products shall transfer to Customer upon VWR's delivery of such Products to the carrier. Shipping and handling fees, special packaging materials (e.g., blue ice), carrier surcharges (including fuel surcharges) and hazardous material fees imposed by government regulation will be added separately to the invoice. Customer acknowledges that VWR may refer to shipping and handling fees on VWR's invoices as "Freight." The shipping and handling fee that VWR charges may not be limited to VWR's actual transportation costs and may include other shipping and handling costs. Products sold through Ward's Science and/or Sergeant Welch may be subject to special shipping and handling fees.

Figure 6-4. "Shipping and handling fees, special packaging materials (e.g., blue ice), carrier surcharges...will be added separately to the invoice."

SHIPPING:

Orders will be shipped within 24 hours of receipt of the order unless otherwise notified. All shipments are sent F.O.B. point of origin. Freight and handling charges are prepaid and added to the invoice as a separate item.

By default all shipments are sent UPS 2nd Day Air Saver.

If you wish to use your own courier account number or specify method of shipment. Please add note to the order comments section on the check out screen.

Figure 6-5. "Freight and handling charges are prepaid and added to the invoice as a separate item."

When I worked in the lab, I hated getting these types of surprise shipping charges. I guess the shipping fee is "nominal" to these suppliers, but an $85 overnight dry ice package for a $20 item is not nominal to most small labs. Besides, if you're aware of shipping costs up front, it may make sense to purchase more of item at one time.

Many researchers never even see these extra fees, since all of the billing often goes through the university's purchasing department, and scientists don't particularly enjoy using valuable research time to chase down invoices.

How We're Fixing This

Buying lab supplies should not be like buying a used car.

We're trying to flip this around and tell researchers what other people are paying.

We put together a list of hundreds of purchases and let researchers have it. Along with connecting with scientists directly, we've also been listing tens of thousands of supplies and chemicals on our website, with pricing and shipping costs.

Using specifications such as the ACS standards, we help researchers find equivalents to help stretch their research grants as far as possible.

I encourage you to visit us at P212121 (*http://store.p212121.com/*). You might be surprised at what you find.

This Is Important

Every year I see great nonprofits raising money to research cures for cancer.

Imagine how much faster science would progress if we were able to cut a lab's expense in half or more.

I've talked with a number of graduate students who spend the first three days of each week synthesizing a compound that their lab can't afford. We've often been able to help them source that compound for less than their starting material. In the big picture, results means scientific progress, which is why we got into this business in the first place.

Science is competitive, from publishing papers to applying for grants and competing for funding. Getting great prices on supplies results in a competitive advantage. There's no reason we shouldn't be bringing this same spirit of competition to saving on our lab supplies.

Sean Seaver is the founder of P212121 (http://store.p212121.com), which was started in 2010 based on his own experience working in a research lab that was running out of funding. Sean received his doctorate in chemistry from the University of Toledo and previously worked at Los Alamos National Lab.

Allison Pieja and the Minor Miracle of Methanotrophs

Glen Martin

Harnessing Bacteria to Generate Earth-Friendly Bioplastics

Methane is a problematic gas. It is expelled in massive quantities by everything from marshes and muskegs to wastewater treatment plants and landfills. And because it has 21 times the heat-retaining qualities of CO_2, it's pretty much public enemy number one when it comes to global warming.

But for Allison Pieja and her colleagues at Bay Area–based Mango Materials, methane is a valuable —indeed, indispensable —resource. The company uses microbes that consume the gas to produce an essential commodity: polyhydroxybutyrate (PHB). Plastics, in other words.

"Our process involves a kind of octopus arrangement of anaerobic reactors aimed at continuous production," says Pieja, Mango Materials' director of technology. "We culture the bacteria in a central tank, and then 'fatten' them with methane in a series of satellite tanks."

The microorganisms, known as methanotrophs, are astoundingly adept at their assigned task. In a methane-rich environment, up to 60% of their mass will manifest as PHB. The polyester-class polymer is harvested as exceedingly small granules from the expired microbes.

But is there any significant value to PHB? For the short answer, simply harken back to the 1967 film *The Graduate,* specifically to the scene where a certain Mr. McGuire is advising young Ben Braddock on a career path: "I just want to say one word to you. Just one word. Are you listening? Plastics."

That observation is as true now as it was 48 years ago, especially for environment-friendly bioplastics like PHB. It's nontoxic and biodegradable and has a wide range of applications, including the microgranules in skin-cleansing compounds, drinking cups, shampoo bottles, sutures, and disposable razors. Prices currently range from $2 to $4 a kilogram.

"PHB is versatile, but not a tremendous amount of it is produced," said Pieja. "It can be expensive to manufacture using standard means. Our process points to both significantly reduced production costs and easy scalability."

And because Mango Materials' production relies on methane, there's no shortage of cheap feedstock sources. In northern California alone, there are enough methane-oozing wastewater treatment plants and landfills to produce 2.5 million pounds of PHB annually.

Numerous attempts have been made to utilize methane on a commercial scale, primarily for energy production. There's a problem, though. Pure methane is odorless, and in its unadulterated state, it is literally natural gas; it could be employed to generate electricity or heat homes. But as its rich scent confirms, the gas that wafts from our landfills, sewage plants, dairy farms, and hog facilities is, well, rather dirty. It's contaminated with compounds that not only make it stink; they make it corrosive, to the point that the gas rapidly fouls transport infrastructure and power-plant components.

"But our bacteria can use methane straight from the landfill or water treatment plant without any effects on growth or polymer production," says Pieja. "Dirty methane isn't an issue for us."

Certainly, anthropogenic methane has vast latent value, even if no one has figured out yet how to turn that potential into lavish profit. A 2013 U.S. Energy Information Administration report (*http://bit.ly/biogas-doc*) estimated that landfills, livestock facilities, wastewater plants, and other commercial and civic sites churn out about 7.9 billion tons of methane a year. That's equivalent to 420 billion cubic feet of natural gas, enough to displace about five percent of the natural gas used in the country's electricity production, or 56 percent of the natural gas consumed by the national transportation fleet. If all that gas were turned into PHB by methanotrophs—well, at minimum, you'd have a heck of a lot of biodegradable shampoo bottles.

You might think that the PHB-producing methanotrophs Mango Materials uses are proprietary bugs, the end result of highly sophisticated biohacking. Guess again. They have been culled, however meticulously, from the wild, assuming you include sewage treatment plants and landfills in your definition of "the wild."

"There's a saying in microbiology that 'everything is everywhere,' and the selection of our bacteria confirms that," Pieja says. "We're not at all philosophically opposed to genetic modification, but we're just not focused on it at this point. We're more concerned about scaling up our cultures with our current strains."

Mango Materials got started when Pieja and company CEO and cofounder Molly Morse were studying at Stanford. They worked closely with environmental engineering and science professor Craig Criddle, who was responsible for the seminal work on harnessing PHB-producing methanotrophs (and who currently serves on the firm's scientific advisory board). Mango Materials was incorporated in 2010 and received its first funding in 2011.

The company stimulated widespread interest and attracted additional funding from the get-go. In 2012, it was awarded the $630,000 grand prize in the Postcode Lottery Green Challenge Business Plan Competition. It received a Phase II National Science Foundation grant in 2013; and last year in Monaco, the management team was presented with an Excellence in Technology Award by Prince Albert II.

Currently, the company is more concerned with refining production techniques than figuring out penetration strategies for the global plastics market. It has a trust agreement with the US Department of Agriculture for the use of lab facilities in Albany, California, and it's running a pilot facility out of a Silicon Valley wastewater treatment plant.

"PHB is very sought after," says Pieja, "but it's hard to quantify the ultimate demand. We have a tremendous amount of interest in this product, (and) we're just trying to keep up with the requests for samples. A pharmaceutical company may only require a small amount, but a plastics firm may need a couple of hundred pounds. In either case, our product has to be competitive with other plastics derived from petrochemical sources. And it has to be easy to employ, which is why we're designing it so it can be used with current manufacturing equipment."

Of course, methanotroph-produced PHB's strongest marketing point may be its transcendent greenness. The process employs a potent—and for all practical purposes, limitless—greenhouse gas as its feedstock, and it harnesses microbes rather than energy-intensive petrochemical plants to generate the product. Moreover, PHB in and of itself can constitute a significant carbon sink. For every kilogram of plastic produced, up to two kilos of "CO_2 equivalent" carbon is sequestered.

Admittedly, that carbon isn't necessarily locked up forever; once a PHB-based cup or bottle ends up, say, floating around in the ocean, it decomposes to its constituent elements and compounds. But just as there are bacteria that can convert methane to PHB, so are there microorganisms that can eat PHB and churn out methane. So Mango Materials isn't just promoting a bioplastic, observes Pieja: the company is promoting an ethic, a virtuous circle of polymer production and recycling predicated on methane and bacteria. Worn-out products made of PHB can be recovered and converted back to methane, which can then be turned back to PHB.

"As long as you complete that loop, you essentially lock up carbon permanently," says Pieja. "We anticipate a minimum 20 percent recovery rate."

So what's down the road for Mango Materials?

"Long-range, of course, we hope to scale up," says Pieja. "We have the bacteria, we have the process, and we've begun evaluating methane producers for our first commercial production plant. Either a landfill or an agricultural facility could qualify. The great thing is that people are coming to us. We're getting lots of requests for samples, lots of people telling us, 'Hey, I have all this methane; help me do something with it.' They're excited about it, and that makes us excited."

Glen Martin covered science and environment for the San Francisco Chronicle for 17 years. His work has appeared in more than 50 magazines and online outlets, including Discover, Science Digest, Wired, The Utne Reader, The Huffington Post, National Wildlife, Audubon, Outside, Men's Journal, and Reader's Digest. His latest book, Game Changer: Animal Rights and the Fate of Africa's Wildlife, was published in 2012 by the University of California Press.

The Present and Bright Future of Synthetic Biology

> *Thus it was that, through the succession of his creatures, the Star Maker advanced from stage to stage in the progress from infantile to mature divinity.*

OLAF STAPLEDON, *STAR MAKER*

Luis Silva

I'm here to talk about the present and the bright future of synthetic biology. As a hacker, serial entrepreneur, and transhumanist, I have in recent years been following and working with exponential technologies that are merging, such as artificial intelligence (AI), robotics, synthetic biology, and software.

We are seeing a growing ecosystem of computational tools, biological techniques, languages, and platforms to program synthetic biology in an easy way. Tools such as Antha (*http://www.antha-lang.org*), SBOL (*http://www.sbolstandard.org*), Clotho (*http://cidarlab.org/clotho/*), Eugene (*http://eugenecad.org*), Merlin (*http://www.merlincad.org*), Genome Compiler (*http://www.genomecompiler.com*), DNA 2.0 Gene Designer (*https://www.dna20.com/resources/genedesigner*), MAGE (*http://wyss.harvard.edu/viewpage/330/*), and Benchling (*https://bench ling.com/*) are allowing researchers to be more efficient and productive.

As a software developer, I compare these tools to the low-level platforms that allowed us to code the firmware and bootloaders of the first computer hardware. But we're using ACTGs and living machines instead of assembly and silicon chips. Some companies like Autodesk want to go one step further, adding a layer of high-level abstraction to the task of engineering synthetic biology, through its new Cyborg platform (*http://bit.ly/cyborg-pro*). Even Microsoft is contributing to the synthetic biology community with a set of tools, languages, and compilers (*http://bit.ly/micro-gen*).

Unfortunately, we still don't have ready-to-go "bio Arduinos" in synthetic biology. If you want to test your code and design, you need to go to a lab and manually pipette microliters of a set of chemicals, bacteria, and genes, following protocols written in notebooks and spreadsheets in a tedious and slow process. I don't need to mention that this work is exhausting and prone to errors. Companies such as Modular Science (*https://www.modularscience.com*) and OpenTrons (*http://www.opentrons.com*) are changing the landscape with plug-and-play pipetting machines and hardware that can automate all the manual pipetting work required to test its bio project. Initiatives like the OpenPCR (*http://openpcr.org*) and Sensa.io (*http://sensa.io*) help you to automate the PCR process and simplify the use of bioreactors.

Some projects are moving all the infrastructure and heavy lifting from a physical lab to the cloud. Companies like Emerald CloudLab (*http://emeraldcloudlab.com*) and Transcriptic (*http://www.transcriptic.com*) allow you to not have to set up and maintain any real workbench to do your research, and Transcriptic is pushing an open source standard to describe experimental protocols called Autoprotocol (*http://autoprotocol.org*).

As the community and the number of projects have been increasing very fast, we've seen the emergence of hubs like Synbiota (*https://synbiota.com*), where you can check on and contribute to a growing number of open science initiatives.

Despite the fact that nowadays it is relatively easy to start in synthetic biology because of the growing community and the tools mentioned, we still are in the early stages when the topic is gene synthesis. The price of gene synthesis has been exponentially decreasing in recent years, and the capacity and quality of the constructs offered by companies like DNA 2.0 (*https://www.dna20.com*) and Cambrian Genomics (*http://cambriangenomics.com*) have been evolving. But the speed of design and code synthetic biology, compared to that of silicon software, is still very slow. After designing and simulating your new organism, you need to order the plasmid sequence and wait days to find out that your construct is not working. Projects like KiloBaser (*http://www.kilobaser.com*) are looking to change this scenario, bringing you a desktop DNA synthesis machine.

As I mentioned before, we still don't have a "bio Arduino". We are struggling to get the first "bio transistors," "bio processors," and "bio-operational systems." Despite all the tools and achievements we have achieved so far, we are taking the first baby steps toward the goal of programming synthetic biology as we program computers. Initiatives like the BioBrick standard (*http://biobricks.org*) and iGEM (*http://www.igem.org*), and inventions like the Transcriptor, are paving the way to the first "bio motherboards" and "bio processors."

When a musician plays an instrument, he doesn't usually know all the physics details involved in the fabrication of that instrument. But even with this "lack of knowledge," he can create beautiful masterpieces. When a software developer creates a tool, she doesn't usually know all the assembly instructions that are running in the processor, nor the details at the transistor level, but even without knowing the hardware internals, she can code amazing pieces of software.

We need more well-designed, high-level abstractions in synthetic biology. Children should be able to experiment with synthetic biology in the same way they experiment with games. Even though technology in the biotech industry is becoming more accessible because of the open source movement, the dropping prices of synthesizing and sequencing DNA, and emergence of DIYbio community labs, it's still very hard for new people to enter the field. We have a lack of curated, organized knowledge and simple, well-designed, high-level abstractions that enable amateurs to create projects but don't limit the productivity of experts in the field. Everything should be as simple as it can be, but not simpler.

By using DNA as a programming language, we are dealing with one of the best-designed programming languages of the universe. It creates its own hardware, and nature has been using a huge library of genetic code in the last billions of years to create practically anything. Let's invite more people to the party, giving them beautiful tools and basic knowledge to express their creativity in a cocreation with nature. With more hackers, makers, and artists inside the synthetic biology community, we are going to see a myriad of unexpected, useful, and beautiful creations.

At Arcturus BioCloud (*http://www.arcturus.io/*), the company I recently cofounded, we are working hard to connect all the dots among hardware, software, wetware, and user experience, closing the gaps in the evolving synthetic biology field, offering to our users a beautiful integrated platform to code life inside a safe and regulated environment, and unleashing the power of nature in an easy way, even for those who are not scientists. Our massively transformative purpose is to bring genetic engineering to the masses. We created a completely autonomous biotech lab and connected it to the cloud, allowing anyone, anywhere, to use it with the assistance of Arc, your AI biotech robot companion.

I'd like to quote parts of a lecture from Freeman Dyson (*http://bit.ly/dyson-vid*), a theoretical physicist and mathematician, that clearly describe the bright future of synthetic biology:

> *There is a close analogy between the Von Neumann's view of computers as large centralised facilities and the public perception of genetic engineering today, as an activity of large pharma-*

ceutical and agribusiness corporations such as Monsanto. The public distrust Monsanto, because Monsanto likes to put genes from poisonous pesticides in the food crops, just as we distrusted Von Neumann's, because Von Neumann's liked to use his computer for design hydrogen bombs. It's likely the genetic engineering will remain unpopular and controversial so long as it remains as a centralised activity in the hands of large corporations ... For technology to become domesticated the next step is to become user friendly.

I talked to the open source crowd about biological sharing. In addition to sharing genome databases biological communities can also share genes. The physical sharing of genes between diverse members of a community gives another meaning to the phrase open source. When genes are shared freely, a biological community reaps the same advantages from sharing genes as the open source community reaps from sharing software. So my fifth heresy says the open source movement maybe recapitulated in a few decades the history of life on Earth over billions of years.

We are moving rapidly into the post darwinian era when species will no longer exist, open source principles will govern exchange of genes and the evolution of life will again be communal.

With a peaceful purpose and the right synthetic biology tools, we have the opportunity to restore our planet and improve the lives of billions of living beings writing a new story about the human civilization. We can capture the excess of carbon dioxide in our atmosphere through genetically engineered organisms and stop the slaughtering of animals to produce leather and meat (Modern Meadow (*http://modernmeadow.com/*)), milk (Muufri (*http://muufri.com/*)), eggs, textiles, and horns (Pembient (*http://signup.pembient.com/*)). We could also produce cheap cancer drugs (Violacein factory (*https://sciencehack.synbiota.com/*)), next-generation biofuels, open source plants (Glowing Plant (*http://www.glowing plant.com/*)) and medicines, beautiful flowers (RevBio (*http://revolutionbio.co/*)), and an infinity of yet-to-be-imagined creations that will solve the biggest problems of humanity and bring joy to our planet.

Luis Silva (http://about.me/luisbebop) is a serial entrepreneur and transhumanist passionate about technology, science, history, yoga, languages, music, art, space, and biotechnology. He is a Singularity University alumnus and founder of an investment holding focused on disruptive technologies, named Merkaba (http://merka.ba/), and sits in the board of a few fintech and healthcare companies. More recently he cofounded and invested in Arcturus BioCloud (http:// www.arcturus.io), the company that is safely bringing genetic engineering to the masses.